When you look up in the sky at a plane, do you think about the history of flight?

Today, we can hop on a plane and go to another country.

Birds can fly. But humans cannot fly. A long time ago, this was a problem. Humans have wanted to fly for a very long time.

Flying machines were built. Balloons and gliders were invented. The need to fly led to the invention of airplanes.

In 1903, the Wright Brothers' airplane was powered and stayed up in the air for more than a few seconds. This changed history.

A few years later, airplanes could stay in flight for more than thirty minutes.

Today, airplanes come with strong jet engines and big propellers and wings. Now, do you want to fly?